FOR CRYING

C000130489

Also by Chris Wallace-Crabbe:

The Amorous Cannibal, 1985
I'm Deadly Serious, 1988

FOR CRYING OUT LOUD

Chris Wallace-Crabbe

Oxford   New York   Melbourne
OXFORD UNIVERSITY PRESS
1990

Oxford University Press, Walton Street, Oxford OX2 6DP

Oxford New York Toronto
Delhi Bombay Calcutta Madras Karachi
Petaling Jaya Singapore Hong Kong Tokyo
Nairobi Dar es Salaam Cape Town
Melbourne Auckland

and associated companies in
Berlin Ibadan

Oxford is a trade mark of Oxford University Press

First published in Oxford Poets
as an Oxford University Press paperback 1990

British Library Cataloguing in Publication Data
Wallace-Crabbe, Chris, 1934–
For crying out loud.
I. Title
821
ISBN 0-19-282789-8

Library of Congress Cataloging-in-Publication Data
Wallace-Crabbe, Chris.
For crying out loud/Chris Wallace-Crabbe.
p. cm.
I. Title.
PR9619.3.W28F58 1990 821—dc20 90-7247
ISBN 0-19-282789-8

Set by Wyvern Typesetting Ltd.
Printed in Great Britain by
J.W. Arrowsmith Ltd, Bristol

for Georgia, Toby and Joshua

*How can we decide which game we are playing? In most cases
we can't decide*   WITTGENSTEIN

ACKNOWLEDGEMENTS

My thanks are due to the editors of the following journals, in which many of these poems were first printed: *The Age, The Australian, The Canberra Times, The Sydney Morning Herald, The Age Monthly Review, Meanjin, Overland, Scripsi, Webber's, Westerly; Antipodes, The Harvard Magazine, Prairie Schooner, The London Magazine, Oxford Poetry, New Oxford Poetry, Poetry Review, Verse; Landfall, Outrider,* and *Sport.*

'The Inheritance' appeared in *The State of the Language*, eds. Christopher Ricks and Leonard Michaels, University of California Press; and 'Puck and Saturn' in *The Foundations of Joy*, Angus & Robertson. 'The Dead Cartesian' was published in *Poetry* Chicago. A number of poems have been heard on the Australian Broadcasting Corporation, on 5UV Adelaide, and on Dial-a-Poem, Boston.

CONTENTS

I

THEY

Where have they gone? Somewhere ahead of us
in a meadow like the square root of minus one —
infinite pastoral; pure interstice —
where two objects can browse in the same space
and history leaves not even a snowflake's print.
They have passed through darkness into a radiance
which we cannot know and they cannot comprehend
but which does not remember the griefs of our world.
The pain is cauterized,
                              the atoms dispersed.
Body is no more body, nor is it soul.
They are now at one with a nearer face of the All.
Lamenting them, we weep for ourselves.

THE LIFE OF IDEAS

i.m. A.F. Davies

Aloe, agave, portulaca, prickly pear,
How these remotely anthropomorphic shapes
Gathered around us in their martian rig
As we walked numbly through an afternoon
Of lessening grief.
                    The eye delighted
In such a weird fair of inflected shapes.

These were the forms I once made up and drew
In backs of exercise books during Latin:
Perhaps their freakishness is vaguely classical.
Rounded up, fed here in a garden of science,
All these pumped-up elbows, clubs and phalloi
Articulate a system clear as Latin

But which I cannot do much more than glimpse,
Reading these plots as barbarous inspiration.
Yes, I'd pillage these monsters for a drawing,
Preferring not to know, in order to
Sneak up on knowing from another tack,
The calligraphic dance of inspiration.

The Life Force has come to a mardi gras,
Stagey bumpkins dressed up in quilted gear
Stuck full of arrows like Sebastian,
Or else ratbag knights from an amateur production
Of *Murder in the Cathedral*, pissed as newts.
How could nature have tricked out all that gear?

Bearing in mind (whatever may be mind) that
Language is the language of languages,
We ought to learn from this taxonomy
Something at least. Such plump allusions to
Plebeian cabarets or rustic orgies
Hide symptoms of the damaged languages.

Damage is where we start from. We survive
Displacing down a value-gradient
The itsy-bitsy fragments of our childhood.
The signified is all that is the case
But what rich slips got buried down the garden?
Grease, grace, gravy, Grandma, gradient . . .

Just naming them will conjure something up,
At best a truth: whatever that may be;
Say, something planted in the numinous
On which the sun leans with particular grace
Like it would on a dancer. What green piercing music
Coerces our shards of happening to *be*?

It is raw grief can lock us into process,
The linkage also grown from the forgetting
As uncollected plants may haunt this rockery.
Some of these succulents are jokes about
Utter non-being —
                    I don't know why I said that.
I feel dreadful. I simply need forgetting.

We build from what we're given early on,
Keeping a child's original box of paints.
Freedom finds room amid taxonomy
And calls renunciation civilized,
Though barmy Nietzsche felt that pride and wit
Could reinvent the colours of the paints.

Language is limited but inexhaustible;
Our bodies are the grounds for metaphor
But forebears tell us how to name our bodies.
The double-bind suckles a double mind
With which we wander through Botanic Gardens
Negotiating them by metaphor.

Names for these shapes?
                                    Why, anthropoid,
Globulous, falgate, serrulated, drab,
Robust, rupestral, pyramidal, plump
And acutifoil. When we get a grip on the names
We take things into our mental block of flats,
Setting them up as crudely green and drab.

In this prodigious picnic of the cacti
Departures from a norm are the grand prize.
A new idea is always faintly monstrous,
Its novelty being what constitutes
The warp or bulge. It is the spanking new
Reversal of an axiom that we prize;

It is the opposite that's good for us,
Taking the dog of habit for a stroll
On the Big Dipper . . .
                                    Heart then trips a beat
Peering into unguessed taxonomies:
Kekulé, dreaming of the benzene ring,
Let fantasy tell reason where to stroll.

Reason, the dream of dreaming centuries,
Listened to what prodigious Sigmund said,
Dwindling back to a tense meniscus: dread
And traumata beat against that skin,
Language papers vainly over the top —
Try it again with a Not in it, you said.

MENTAL EVENTS

The brain,
a fat grey flower,
blooms
    on a stem
        of bone.

The colours it desires
are unbounded
tropical
    tasteless
        fauve.

It feeds
on darkness
like a flock
    of vampires
        sucking away.

The whitecollar bosses
require it
to shake hands
    with a terminal.

'Up you,'
it cries,
flouncing away
    on clouds
        of homespun gold.

Bouffant, suave, grouse,
orchestral music
by Handel
    skirls through
        the tossing hair.

THE CONCEPT OF MIND

Where is it
and how big?
Does it have curved walls
or else particleboard compartments
with a dancehall for the past?

Is it like
a suspender belt?
Are its doors the eyes of a tree?
Has it room for resurrection pie?
All the maps have passed it by.

Is it like
a blue wren
or a body and a ball?
More the shape of a hectare of landscape,
largely humdrum run-of-the mill
but with creekbed loams —

yet perhaps
it is more
like team spirit or the C. of E.
and hardly hydraulic at all,
so my thoughts don't call for wings

and won't fall.
But suppose
that modestly mousy mind
gave up on point of view,
what immodest expansion!

It would be
the whole bouquet
of Ego's red letter day,
demanding equality
with the huge mailorder catalogue:

all of those
leaf-thrashed greens
and hyperbolical oceans,
throng, brick, plain and grain.
What if they went fifty-fifty?

Mind on tiptoe
in the left-hand scale
and world like an old balloon
squeezed rubbery into the right
while Justice kept her eye on the needle.

having herself a ball.

But anticyclonic
winds are gusting
and the scales aren't there at all,
or so I think in a dull blue-grey
when I get to think about thought.

How does it work, then,
flat, slant or vertical?
Mind is a close call.

DOMAIN ROAD

Delph, our tangled spaniel, bounces
briefly up a steep backyard
and I am four years old, under leaves . . .
Where did that slice come from?

Somewhere inside my being
the enzyme calpain has played a part,
for every performative occasion
making a collage of neurons.

The dendrites frond like little trees.
My brain is a kind of Metro:
it would be really nice to have a ticket
to see me round and home.

A neuron accumulates its crop of signals
then fires a small gun.
Axon to synapse the signal goes,
molecules pushing out from their dark shore.

Receptors flush, the neurons are busy as bees.
Oh dear, none of that chubby frame
is left in me, all new process
fifty years on

but in what vein has that scene been buried,
nectarine tree and terraced carrots
with Delph panting on the back steps?
Channels are opening in the dendritic spine.

THE INFILTRATORS

for Graham Little

And if your supermarket trolley
Keeps tugging quite inexorably
Off to the right; or some nong dents the car
While you were off buying the vegies,
You know you're in the grip
Of an undertow
And it's never going to let up on you now,
Not even when, chattering, say,
To some coruscating young mind
You reckon you're tapping a Life Force
Straight from the mains.
                                        Stiff cheese, mate,
Some bugger somewhere is writing out the bill
On tablets of scar tissue.

Crafty Rilke twigged that the future
Begins to shape itself inside us
Long before it happens,
Like lumbago.

                        Just remember this,
Tendrils of the next generation
Are circling serpentinely
Or pale as shades
While they go tracking all about
And in between us, here ensuring
That this year's radical lads
Are made over by dark fiat
Into next season's boring
Redfaced old blimps.
Listen, then,
                        ear to the ground:
Those new peagreen beliefs
Are beginning to sprout,
Earth crumbling back off them
As they take to the thin air.

It is only by breaking
The necks off old bottles
That we can drink the dark shiraz
Of not remembering.

THE EVOLUTION OF TEARS

I wonder whether grief was already invented
as far back as the Mindel and Würm glaciations,
some rogue gene having tipped the scale.
                                    I wonder
whether the human creature was chosen by grief
as much as by stone tools or opposable thumb.
Did our hopeless remote forebear sit blubbing his eyes out
(her eyes?) by gunyah or cave? We cannot know.
Tears leave no grooves in archaeological sites,
a broken heart has never been trowelled up;
the *lacrimae rerum* do not resemble objects
though misery be as hard as a stone in your hand.

A long way down the line were those other properties,
the lily of logic with a rose between her teeth,
rationalization, debt, the barque of state,
but darker human traits had crept in first
teaching the dread monosyllables, lost and gone.
One, more obscene, begins with a capital D.

Chemical? We comprehend grief, but not always.
A burial rips the guts out of everyone close.
We fail. We suffer. One slides into the ground
who had been a spirit, horsed round and laughed as we do
and now joins the majority . . .
                                    the tacit ones . . .
in a horizontal kingdom underfoot.
We do not understand.
                        Together we rise,
feather, turn and fly away.

RIVER RUN

for Kevin Hart

In the beginning was John.
He had shares in the logos
and could treat all that stuff,
antimatter and magic,
with elastic familiarity.

But the name of the legion was legion
so that stocks grew bearish,
world swam under thick waves
of coarsening history,
and the pi-meson said nothing.

A small ache behind my clavicle
may be arthritis,
a bend in the third world
unpick itself as torture,
the inexpressible word.

We puff along like beetroot,
a sweat of scullers following
their coach on his bicycle.
Can I parse the figures of eternity
with my shoulders between my knees?

GRACE

Summer has peaked for now,
say the dense but extroverted
magenta clusters that angelize
our crooked flowering-gum
tempting bees to sip and die,

while the taste of mango segments
begets no language:
it is like mature Chopin,
diving into some chill rock-pool
or sex in a classic novel.

Summer begins growling
(a geriatric politician),
its big heat suddenly falling
in the guise of sixpenny raindrops.
Hang on, says the thunder,

scarving the swallowed east
with purple-grey.
                But the sun bangs back.
In this wide, orchestrated glare
soul fully inhabits the body.

GYPSIES

The north wind mutters to me
there are nooks and fissures
in racial consciousness
where deviant flora dwell,
burgeoning only after decades.

Under those thin gums
at thistled Kororoit Creek
the Gypsies used to camp
with their tents, dogs and cars
when I was just a kid.

Of course we drove past them:
one is always driving
past something without concern
until it proves too late
to record the social wound.

Once I saw them close up
in a closed Plymouth sedan
getting petrol at Seymour, I think.
Dad reckoned that their queen
lurked behind drawn blinds

for some ritual or reason.
Suddenly I don't even know
whether Gypsies flourish at all
under the Southern Cross.
nor where the clean past has gone.

MALTHOUSE

In the dry kiln
they turn the turning barley
with near-square wooden shovels.

Their feet
are lapped in prickly hessian
against this heat.

They inhale dust.
Bending Tolstoian reapers,
they flip the shine of golden grain,

backs licked with sweat
like vaseline.
They are ribroasts inside a stove.

The barley aspires to malt
and will slither down a tin shute
for bagging. For beer.

They take a breather.

HIGH RISE

Ah, but all those lamps are computer games
on a big townplanner's dazzling scale
or Mercator's map of the mind.

In America it is commonly the case
that the first gin-and-tonic does the damage
so that the wit of a skinny professor
sags in perspective
perfecting a cardboard role
in which there can be no salvation
except for downright size
which is at heart a spiritual quality
like bounding muscle or intelligence,
or the peace of god which baffles understanding
and smells of burning gumleaves in the grate.

But the loving aerogrammes have failed to come,
brittle and blue
with news of Airey's Inlet, Lygon Street,
old footy scores and the like,
our only letters boasting business windows
as I stare from nineteenth storey windows
at a romanesque powerhouse in retirement,
angled highways by still waters,
lights bedazzling like a jeweller's shop
and baby aeroplanes here homing slow
while the reflective river just winks Yes,
for which my thanks
until the muffled stars yawn and turn over
in a warm susurrus of nocturnes.

Oh, really, have you been here a long time?
Only as long as my arm . . .
and the clock in its white tower strokes midnight
like the feathers of a bird
ruffling against the cold rhetorical wind.
Oh yes, a long time now.

THE SOUND

'Hush,' you murmured, 'listen.'
And we cocked our heads in a burglar-watching way
Until you decided the coast was clear.
'It's a sound like little chickens hatching,'
You then concluded, made your way upstairs.
I looked out the back window, through the fuchsia,
Toward the overbright stars
And the whole sky was full of chickens falling,
Plump as partridges, gently bobbing down
Or floating over the garage like balloons.

Taking it very much as it comes
Or with a pinch of salt,
Such a fluffy epiphany of chooks
Must have meant something particular after all.

It was a sound like little chickens hatching.

MY SURVIVING SONS

Stars flaunt quite strange again.
My sturdy vulnerable
tanned little men

swagger a little
having now lived in three
continents before rounding nine.

Eighteen floors off earth's rind
the lounge windows are paintings of Cambridge,
the bedroom, pot-and-kettle city

and life has frayed out
like an old rug shaken over
a silver wrinkled elbow of the Charles:

I want to buy more books
brightly to prop me up
before we kick a football round the park.

I want to cry in the idiom of blood
sobbing away through a pulse
in earth's wrist

or to sing like a dawnlit cloud.
Flaunt, ambulant mortality,
golden delicious apples on your tree.

Fluently, dearest wife,
tuck me under your wing
while seagulls float around our haunt like knives.

AN ELEGY

Everything turns out more terrible
than they had said, or what I thought
at midnight they had said,

but the dark marks
tracking across clean snow
way down there must be people,

that is
if anything on earth can be human
when eighteen storeys below,

so that I wish again
it were possible to pluck my son
out of dawn's moist air

by the pylon-legs
in that dewy-green slurred valley
before he ever hit the ground,

to sweep under his plunge
like a pink-tinged angel
and gather him gasping back into this life.

LOSS

Experience was only
what you lose every day,
huge blown-away clouds
which memory may

think to have drawn
back live to you
but those images
are all untrue.

The only trick
is to write them out
replacing dead life
with phrases about

what verdantly might
have happened then
till amber clouds float up
from your table again.

WORLD

Alas, the geographical cities
up and drifted away

in their bangles and ganglia
of meretricious lights

ingenious as a cat-scan
or bleeding computer game,

their patterned haemorrhages
turning the black air on.

They emerge as nameless monitors
of blear-eyed self,

the argument from design
here abacussed in lukewarm freckles.

Looking down, I can see
a giant's irradiated cell-configuration

pinned out just for me.

But I am drifting out of all reach
like thistledown.

FOR CRYING OUT LOUD

Here
quite as much as there
in the dead straight street
or snailed and breathing gardenplot

time,
that sarcastic medium,
ran silvery through my fingers
like sand, or bonemarrow.

It
leaches through every life
which steps gingerly into it
under and over again.

We
were set down on moist earth
as though to train for some Grand Final
which is never going to take place.

Such
living as you greenly had
flowered and fruited boldly after all
but you misread what it meant.

Now
there's a torrent in the blood,
a sense of arabesques fanning out
across their shining enormous mudbrown delta.

In
the sepia vision, daily, diagonally,
you are walking in mufti backward
through yourself.

CONCERNING AN AFTERNOON
IN NEW YORK

How strange it all is.
Overbright
flat
sun hits
the pates and shoulders
of high buildings
like a gong,
bleaches out
the sticklike trees
and hauls forward
this or that
random pallid facade
on the Jersey shore.

The sky
is an indiscriminately
unreverberating
cyclorama
of pure dust blue
a long way off.

Ah, really?
Fashioned like the letters
A and R,
people far below
toddle to work or jaywalk
across pale macadam
which has been marked out
by somebody or other
with bonewhite Ls
or dashes.

Cars
which are not like letters
snail, behave and crawl
according to their lights.

OLD TALES: MOSCOW

In Friendship House
light stanchions like mailed arms
reached heavily out from the wall

and heavy heads
nodded above the tea and fruit
bobbing under the slippage
of leaden wartime memories.

There ought to be
a nook for God in this narrative
but only musical theologians
know where it should be.
There have been times,
deep green times,
when the answer was just as plain
as a cup of tea.

Today was different.
I waited behind the scalloped towers
for Beast to come panting in
and delicately court his Beauty.

Hush!
The wooden squirrel freezes
on his bough.
And the stag dreams.

THE IBIS

Morning. No matter which clod we voted for
politicians will go back to Canberra;
you can't replace them with goats.

But here, in the serene glass,
hairy tussocks are wearing rhinestones
and the sun appears hunched

behind a strip of pewter cloudbank
while the big moon-face
sits on an old rooftree, hoping to set.

But here come the ibis pulsing back
in vees, in skeins, virgules, diagonals,
their rushing sweep flooding my head

with the recognition
that these are the gods of a writer,
calligraphic, fluent, with heads like pens,

and they complicate the parchment dome with signs.
They have scrawled across it, 'Live.'
They swim down into mere west.

INVERSION LAYER: OXFORDSHIRE

On one tilt of a silt-valley dulled with haze
I turn the large key, *ker-chunk*:
strange stars blink pauper-pale
while a billion dozy leaves murmur, 'Really?'

I believe I can understand
the harvest moon through its plummy accent
but nature's great book might here as well be printed
in Swahili.

Really?
      Well, yes and no,
this life trails after obsolete forms of art,
prettified clouds mound like watercolours,
the birds vaunt etymology

and, come morning, us still alive,
middle distance will tuft with amber and brass,
copper, orange, the tardy spatulate green:
technicolour versions of fall.

Did I fly Economy all this way,
tailfeathers dropping into several seas,
merely to squinny, take notes and try to cure
a lifetime's hatred of gothic

or, worse, of framed architectural prints?
Beggared if I know, but a wind from Wales rises
you can hear the Green Man snap tops off trees,
rustling his vicious cloak and crushing squirrels.

'I'm a non-believer,' I whisper.
'Be off. Go stand in a fairy ring,
disturb the filthy starlings
and may dry rot powder your corky soul.'

THE INHERITANCE

Dunked into life, a squalling brat
apeing the role of perfect child,
I let this language buoy me up,
shock troops lightly graduated:
nasty, nice, nectarine, nasturtium, noun.

The stuff was rich as mother's milk.
I couldn't see it didn't fit,
making it do so anyway,
eliding what was grossly wrong.
Origins prove nothing, said William James.

I romped round discourse in my room
only devouring foreign books —
northern, that is — containing heath,
lorries, wolves, bobbies and snow.
The signified was quite inadequate,

a mere Australia. City fathers
had long conspired with Empirespeak
by cancelling native foliage;
so every winter English buds
flashed into fluffy pastel bloom again.

As cunning as a leaning dunny,
this international currency
parades its virtue in old rhymes,
tomb after tomb, as death with breath.
We swim along with it. We swim and drown.

THE COLOURS AND THE
CONSOLATIONS

Rolling, aflame, kapoky, stacked,
the clouds have been delicately punctuated
by downslant of a DC9.

They fluff behind the cradled pate
of a crafty historian in blue,
zonked out by some nong's crazy paper.

A cyclorama of eiderdown
or malleable marble drifts
over itself into

tangerine, raspberry, blush, flamingo,
of which the shadows' name is blue
or even a deep purple . . .

These/those
        dozy mounding clouds,
                all such
extravagant
        selfexposure
                of cumulus
coloured
        like old sundaes
                from Ernest Hillier,
wash
        this campus
                into the strange one
        that welcomed me
        in fifty-three
                (exhilarating traumata!).
The voice of English murmured,
        Eliot,
                Bronte
(a bracelet of grey hair about the Donne)
but social sciences
        roared back,
            AUDEN.

                              Look, strangers,
out of this   ̄
                 gaunt academy now
Sydney is sinking
                 wet in the dry economy,
but a paid airman
                 will see me
home.
              The vigorous shadows dwindle,
                         the cheeks of cloud
bunch darker
                         until
we slide sideways out of our names
from hawk to handsaw.

THE BUSH

for Seamus Heaney

Overture:
              violins:
it is all scraggy,
wideawake,
                 ironical,
decked out
              in denim fatigues.
Witty and welcoming,
                 leathery-evergreen,
bemedalled with beercans,
cowpat and wallaby-dung,
flap,
     nub,
         hinge,
             node,
blindeye quartzite,
                 wafery sandstone,
bright as a button
subtle for mile on mile
far from vulgarity
                (far from sleek Europe)
in its array of
              furniture tonings
sheeted by sunglaze
              lovingly dusted,
wispy and splintery,
tussocky,
             corduroy,
all of its idiom
dry as a thesis
to moist outsiders:
wonderfully eloquent
              on its home ground,
branchful of adverbs,
lovingly
             wombat-hued,

dreamily
                    sheeptoned,
fluted with scalloping surf
and every step a joke.

BANKSIAS

for Evan Jones

Handsome, their trunks rise heavily
corkish, grainy, magisterial
above the tilt of dunes;

their leafage fretted against the sky
with a musical delicacy,
undersides oddly white

as the clouds of themselves.
Cylinders of bronze
or lemon fur bedizen them,

turning in time
to childhood's Banksia Men,
coffee-dark, grotesque, multibeaked.

They adore the dovegrey sand
sharing with it a love
for various kinds of dry discrimination.

They commemorate Sir Joseph with a flair.
They represent a primitive brain, or else
the merely picturesque.

A SUMMER IDENTITY

for Jonathan Aaron

Densely fledged hairdos
on dark flowering gums
have put on their habit:
scarlet as revolution.

One of these fine days
we shall all be somebody else,
all the colour of sunlight
crisply at ease in our roles

and the doomed economy
will by then have turned out moist.
Small cumulus clouds are rolling
out of the west like soccer balls.

Down in the bluestone square
someone who stands for something blue
is barking pacific imperatives
through a loud-hailer;

on light breeze, the big sounds
dip and bleat
wavering out over
those tufts of arrogant blossom.

TWO FRUITS

It is everything
in the mouth of poetry at once:
it is creamy Shakespeare.

Glossy black pips
lodged in the custard apple like trouvailles
are perfect where they are,

slip off your tongue
as ecstasy goes memorably down
the little red lane.

Nothing on any tree can possibly trump
this fruit of angels,
pure epitome.

Whoever it was that used
to call this wonder bullock's heart
belonged to a different frame,

a prose cafe.

But the mango,
the arrogant imperial mango
roundly replies
out of the heart of the sun,

I am the only fruit
whose colour
turns out to be
exactly the same as his flavour.
In me you will observe
the dangling miracle
of cognition.

THE DEAD CARTESIAN

Part of me flies upward
or at least elsewhere,

thiswhere being the damp location
of newly separated body.

Thought is not going on
in my current neck of the woods,

my present nook of brown wood,
consciousness this blue afternoon

discreetly redistributed.
Mourners have chunks of it,

much has gone bobbing away
and settled like strings of lint

in unregarded corners.
A trickster god vacuums them up.

Words in their newspaper column
may have been what I am now,

an atomic thought in the mind
of overweening matter.

A SIGHT FOR SORE EYES

Goggles, lorgnettes, four eyes, pince-nez
or, in the old saw, men never make passes,
et cetera. We grow to need them more
and sputter, 'Where have I left my glasses?'

Marvellous extensions of the marvelling self
de Spina of Pisa invented for poor blind men,
they give us a new innings in space and time,
sharpening us back from Monet's world again

which may have been charming, but you missed your train
and couldn't spot the little streets in Melway's.
We need to get our priorities right:
print, instructions, literature, railways,

you can muck through the years getting them straight.
Al Hazen, the hookah-puffing Moor,
picked up the magnifying power of glass
and Roger Bacon saw what to use it for.

A century later we discern,
supple and powerful, Cardinal Ugone
in a church fresco, with riveted specs,
two lenses round as macaroni.

Glare, *Glas*, glazing, glaces,
all allude to this fire-forged ice;
Narcissus knew the eye inflames the ego;
in classic optics, Ptolemy is wise

and Archimedes turns out burning-glasses,
light and music unravelling the ground
for much of science. We suffer and learn
something from light, magic from sound.

We get lost in a cold, crystal world
but our life knows where we are,
shuffling embarrassed across the room,
and a mirror sees how far.

A kid, I looked in a smudgy glass
and now have put aside
my toys, my children's books, false hope.
For this, Galileo died.

AND THE WORLD WAS CALM

Sandbags of sugar cannot conceal the gloomy fact
That we are inserted headlong into life
As a new pen is dipped in lavender ink.
We take up a space amid the comings and goings
Haphazardly, wanly. Velvet wrappings of eternal night
Contain our small blink, pitiless in their Logos.
Powderblue through gentle distance, lyrical mountains
Look at our passing span with incomprehension:
We never read their huge minds, and we die.

Why was the serpent given access to language and stuff?
Awareness becomes a different kettle of fish when eked
Out in a long line: the clipped image gives over
Until your modulation from rambunctiousness to grief
Is felt as a matter of slow brown flow, all river
And no cute islands. Remember, grace yields to Valium
Down on this late-in-the-century flood plain
Where even the fodder and grain crops are postmodern;
That is to say, containing no vitamins.

I'd like to build a cabin or humpy for awareness,
Something rustic like the Bothie of Tober-na-Vuolich
Where I could hum and tootle against the wind,
That long grey stranger skirling through everything.
Grammar is always complete, but the world is not,
Shrugging and folding, surely hatching out of magma.
'You see what you are dreaming, but not with your eyes',
Said a chap who met little sailors down in the park
But grew austere as the case went on wearing him down.

Light is more mysterious than anything else —
That is, except for having to shit and for loving,
Categories you could not have invented, supposing
You were God for a while.
                         Am I alive or dead?
A question basalt or sandstone could never ask.
Awareness becomes a different kettle of herrings
When it's applied to the psyche of a whole country,
Something quite rosy, hydra-headed and fat:
Public opinion did a jig on the carpet of madness.

In the beginning green verbs went bobbing in space
Which was pearly or golden in its painterly turn
And we do not think about gales in the Garden of Eden
Nor about any distinction between plants and weeds
So that Adam is constantly doing something with roses.
Rubric, baldric, erotic, I brood on these terms,
He could have reflected, leaving his pruning aside
While he rolled the well-made words on his tongue like stone:
For the main thing then was learning how to think crudely.

Subtle as ivory handles, we think we are now
Umpteen years on from that scene in the Olduvai Gorge,
Outside the gates of which St Michael struts with his sword.
Pining, we find little paragraphs in which to lament
That we are inserted headlong into life;
Poetry survives with its coppery glint of gnosis
Along one edge.
                    It is a drug that endures
Riding atop the bubbles of evanescence.
The river we step in will burn us off at the knees.

II

To be a character,
that would really be something.

TWO SPIRITS

Mascara running down her face
the sybil crooked her hands and shrieked,
'All of your dreams
are nothing more than torn pages
from the one over-arching novel
which not even I can read.'
In her sacred cage she rocked and shrieked,
'All of us are demented critics.'

Maddened, rampaging through her grove
musky with dust and eucalyptus,
the muse of lyric murmured back,
'Such leaves and waters
are the bleeding grammar of persuasion
which not even I can staunch.
It swims under all of us.'

-

PUCK AND SATURN

No more than this: the buoyancy of the world
first fed me home. The bounce and jump of it;
carolling like a magpie, spry
as the twinkled air,
adrift, alone
on cushions of sweet darkness
I come by.
Tricked out with stardust
I bedevil you,
spanking through gossamer revolutions,
dreamer, blockhead, faintheart,
more than you ever know.

Take note of this, then,
but do not say a word
if cicadas drill through me,
sap rises in wrist and ankle,
air throbs,
time is detachable,
dry leaflets drift down
onto my chest-hairs,
bowels yearn for a nameless harmony
and the full moon bobs fatly,
stabbed through by she-oak needles
just for me. These bones
are all made of rock.

PUCK DISEMBARKS

That sun is glazing and glaring from the wrong direction.
In his government regulation gear
And cultural arsy-turvitude
Puck steps ashore in a grammar of ti-tree.
    He rocks the pinnace.
    The foliage looks pretty crook.

Even a spirit can fail to be gruntled
Standing on his northern hemisphere head
In a wilderness without fairies or dairies,
Whose Dreaming he cannot read.
    He tweaks a tar's pigtail.
    This land is all wombat-shit.

The mosquitoes lead him to think of swallows,
The dipping swallows of Devon
And these alien magpies can sing like Titania
In love with a kangaroo.
    Puck waters the gin,
    Peddling the balance to snubnosed Abos.

The glittering wavelets throw on yellow sand
Big shells like Wedgwood ware
As the imp rises inside him, getting ready
To rewrite Empire as larrikin culture.
    He daubs a first graffito on
    The commissary tent, GEORGE THE TURD.

Against the pale enamel sky
Rebel cockatoos are screaming
New versions of pleasure:
This is the paradise of Schadenfreude.
    He begins to adore
    The willy-wagtail's flirting pirouettes.

NOSE

The running process
with her purposeful cogwheels
makes a point
of every poor thing,

even, say,
the indelicate nose
for all its lack
of iconic power.

Ah yes, or alas,
we all need noses
to recognize
classical beauty by

while humble mucus
telegraphs the cortex
with a wildly seductive
idiom of flowers.

Hmm, but the shape
is in no way noble,
all the way from aquiline
to Susan Hampshire;

beak, blob or smidgen
in various colours,
telangiectatic
at a boozy worst,

I give you the beak
(free now from hay-fever)
sniffing a table wine
or drunken rose.

A BIT OF SHUTEYE

While thought grows mothy
with late fatigue
jazzed-up images —
perhaps in league —

flick through a theatre
of commedia dell'arte,
quirks of queer phrases
joining the party:

spavined proverbs
and theological wit,
micro-Dada speeches,
all do their bit

as midnight leaches grammar
out of my head.
Whoever speaks me tells me,
Piss off to bed.

When thought grows blurry
with feathers and flock
the dreambird will feather
each passing shock.

THE MAN WHO COULD LOOK LIKE ANYTHING IN THE WORLD

He walked out in the dewdrop morning
looking like three million dollars.

He strolled out after lunch,
and he did.

A few drinks late in the arvo
and then he resembled the world itself.

A good long yarn at glinting sundown,
and he did.

Over dinner
he looked just like a diner;

over coffee and some Haydn
he faded into the armchair upholstery.

Then he became
one of his favourite books

and all night long
a buccaneer-scholar-sportsman in cream flannels.

At breakfast, for all the birdcalls,
he looked his age.

After shaving, he looked like Mata Hari
slinking along with his leather briefcase to work.

His bruise-purple shadow
kept a blind eye on him.

SEALING MEN

for Peter Pierce

Grey colloping waves that rasp
Along these halfmoon beaches
Emblematize eternity
And all the lives it leaches,

Worst and most vividly those hoons,
Grogrotten Bass Strait sealers
Laying the violence of the gale
On Aboriginal sheilas;

For those who threw a shanty up
At Portland or on Flinders
Could dine like lords on mutton birds
Baked rudely in the cinders.

Though history swears they laid their blight
On Adam's darker daughters
There must have been some Christian souls
Living off these brisk waters,

There must have been some ferreting
Reflective mind here, surely,
Reading rock, she-oak, cunjevoi
For nature's wisdom purely.

THE LAST RIDE

i.m. Vincent Buckley

It was on a sabbath morning
A hundred years ago
When the good folk of Glen Eathie
Had all gone off to kirk,

All except a herdboy
And his minikin sister
Who were lounging in the heavy grass
Beside a cottage wall:

As the shadow of a sundial
Lay on the line of noon
They saw a motley cavalcade
Wind from a woody hollow,

Weave to and fro round bush and knoll,
Turn by the cottage gable
Where the children crouched in fear,
And slope away to southward.

The horses were shaggy, tiny things
Bespeckled grey and dun;
The riders crooked, wee, bizarre,
All dressed in antique jerkins

Of plaid under their long grey cloaks
And crimson caps which barely
Restrained the tangles of their hair:
Each so uncouth and dwarfish.

The boy and his sister stared amazed
As rider after rider
Passed by the cottage dumb as stones
And wound into the brushwood.

Not till the rout was all but gone
Did that herdboy find his tongue:
'Then, little mannie, what are ye
And where is it you're going?'

The last one turned in his saddle then,
'Not of the race of Adam.
The People of Peace shall never more
Be seen on the braes of Scotland.'

AIR FORCE, BURMA, 1942

Green as a lukewarm salad,
Rangoon; my chum the major,
fluent in lots of queer tongues,
has them too bluffed ever to
try any pilfering now . . .
and then they economize
by beating sly vendors down.
(This place is like an oven
after the high Burma Road.)
And how glad the houseboy is
to have no memsahib around:
'Gentlemen, learned hakims,
mems that illustrious men
in moments of unwisdom
chose to take on as spouses
get nervy in this green heat.'

I sleep with loaded pistol
under my sweat-wet pillow:
so far no night marauders
except for just one who stole
Iqbal's puggaree, also
one civet, dogs, a cobra.
On a fruitripe night we've lots
of crawly, jumpy, bumpy
critters all over our walls;
those lizards look transparent.
The month of oranges
will be with us pretty soon
if the Japs don't blast us first.
An apricot moon creeps up
into the opaque blue sky
like an omen of bombers.
They will fly in soon enough.
This waiting brims up with fear.

Schwebo: faces yellowed by
such doses of atebrin,
it seems that we never have
been alive anywhere else
virulent greens become us;
temple-band and crash of bombs
have always been our music
and the Japs our one true love.

I have to think about Mae
but the thought proves disturbing.
She is flatly beautiful,
cleans up, shares my bed, poses
for portrait snaps—and she cooks,
but it's all got confusing.
Her parents: what can they think?
I asked her how old she was
and (Christ!) I think she said twelve,
so I thrust it out of mind.

The way to survive out here
is. . . Never think of Hobart
and keep yourself in fair trim.

Rangoon? It fell long ago.
Mandalay, city of kings,
has gone, and Mae is now gone
back into swallowed strangeness.
Peace has died, with excitement
since highwire storms fade fastest:
in jungle, memory rots
so that my self has become
a ghost-soul at plod across
interminable ridges,
small tribes with their pipes, bamboo
crammed thick as history,
tropical sores, mud, the shits.
Without Iqbal I'd go mad.
It was grand the other day
when some Chin levies tacked by,
at ease in this steambath despite
old gats and tartan blankets.

There is no colour on earth
I hate more than pale slimegreen.
My knees and my poor arsehole
have born the brunt of this war,
a mind in a skeleton,
part of the dulling skeins
of a ruined army, bushed —
and sweat is my middle name.
My soul shrank to a black hole
deep as these fucking gorges
we have to inch over on
odd rickety, lath-and-rope,
sky-high bridges. But you learn. . .
what do you learn?. . . oh, nothing,
no more than Mae ever learned.
Who can ever know someone
else at all?

       Look, a Gurkha's
corpse blown up like a puffball

Look, this is Burma.
                Listen
that faint 'Boum' was history.
Say, water's my element
but earth's my fate — plus insects —
and the doom they've offered me
(if there's yet some *me* inside)
is threading horror-salad
with Iqbal and suchlike wrecks,
as reliable as pus,
trudging towards India
for ever,
           yes, forever.

THE OTWAYS

'man is no fixed type, but a transitory product'
BRENNAN

He scrambled down the last flank
through prickly-moses or frail
fringes, headhigh bluegums
bent by years of salt bluster,
tussock grass in its bunchings,
dry, uneven, slippery,
aromas in moist gullies.
He knew it all now by heart;
these mountains were his body,
their perfume of musk and rot
could well have been his past.

Heat, salt: a deceptive calm
reigned on the rolled hills like haze.
Kicking out a shower of stones
an echidna bulldozed fast
fleeing some nameless danger.
One cloud lay on the sea;
decades drifted on those tides
rippled and crimped as a brain.

Caught in a brambled creekbed
under leather leaves, over
rapidly washed ellipsoids
there lie the *ad hoc* remains
of a deserted railway,
tramway rather, rust-raddled
and twisted out of all shape.
It used to bring down timber
to a nonexistent pier:
such iron may be called passing.

Drew Cameron leaned over
his pinegreen balcony rail,
stared over rosy gumtips
and the small beach far below,
Finn's tiny figure walking
and the age-old voice of sea.
It was all one harmony.
How many on this high, groined,
magical coast, he wondered,
are living the Secret Life?
What do we know of them all?

The point of noon yawns always
out of reach, raw paradise,
and we cannot feel the names
our tongues have laid on our lives
or on this leaf-thrashed shoreline.

The murderer rattled down
through unfamiliar clumps of
ragged brush, messmate, she-oak,
self: nothing at all.
A rich mix of aromas
outflanked him but stayed blank.
Mind raced quite elsewhere; body
was moving somewhere else, too,
where *her* body slumped like clothes
near agapanthus clumps
in an oldfashioned frontyard
lairy with heaped nasturtiums.

At the end of the Lower Cretaceous these deposits of mudstone, shale
and arkose began to crumple and lift. A template of the Otways was
taking its form. But merely to think of those millions of years passed by
engenders a terrible melancholy: one's own trivial shiver of dread. And
the Eastern View coal measures take their turn to rebuke us.

Flexed under the dread of space
Finn was on a halfmoon beach,
striding over its whiteness
hard against the upper edge
of rhyming waves: not bad surf
though creaming just too far out,
he thought. And a fear flicked through
some parts of him abruptly:
why had he left Annie there
by herself, on the spooked ridge?
They had feared nothing before
but summer fires.
                The chill passed.
He picked up a nautilus,
lightly broken, lovely still,
treasure for contemplation.
Yes, there is this to be said
about terror: it passes.

Knowing what dumps Todd was in
and the dark of his temper,
Marina was gunning fast,
her yellow coupé clipping
round those tight bends like crazy,
kissing gravel and flicking
lower tips of drab leafage.
Butterflies drifted upwards
in wafts of sunlight; moths, too,
and a wallaby thumped off
to wherever safety lies.

Ice-cream wrappers blew against
the tyres of surfie wagons;
the morning paper bloated
with news of an election,
or not, as the case may be.
Wind skirled through the legs of gulls.

Cagey, reaching the beach road,
the murderer nipped across
and into clifftop cover,
zigzagging from there on down
the crumbling near-cliff, puffed hard
and came out onto the rocks.
Not a soul. No-one at all.
Helpless as a bicycle
he slumped, tipped forward and thrust
wicked hands in salt water.
Stones rolled in their small basin.

Before its enormous quest
for unity of a world
the psyche breaks up like rain
yearning, dispersed and falling
but able to remember
the gone angelic chorus,
Debussy stirred with Mozart
on the back of the mind's tongue;
frail futures drum in our bones
in shattered arpeggios.
They really happen.
                    Or not.

When Marina rattled up
the bluemetal drive under
a rolling scent of bluegums
Todd was happy as larry
painting the roof of the shed.
No ragged gloom weighed him down;
he poured a couple of beers.

Finn carried the nautilus
over the soft dune and up
the narrow track to 'Four Winds'.
Annie came out in shirtsleeves,

gasped at the frail nautilus,
hugged Finn, was all freckled smiles.
Teasingly, 'You're just in time
to fillet the three flathead
we're going to have for tea.'
A sudden uproar outside
signalled the kookaburras
winging in for twilight scraps.

Like a close-up shot of hell
the blowhole churned and gurgled,
its broken waters making
a little anthology
of changing, consistent forms.
It is sheer despair which grinds
beneath our glimpse of beauty.

Never no more, the glaucous waves and serried ridges reply in parallel
series. Photons tremble. Tectonic plates are budging. And the stars
keep throwing down their little spears. In the short run, our medium
calls itself time, somehow justifying a promenade of deaths. Never no
more, whisper a billion scented leaves on the coastal gums.

Climbing back up spongy loess
the murderer knew he must
front his music after all.
He thought of her small body.
He envisaged where it lay
and how long it had slumped there.
Rehearsing words for the cops...
something luridly struck him
like a fire across dry mind:
he had quite simply never
killed anybody at all;
pure, undistributed guilt
had driven him down the cliff
to the liquid and clear dark,
to that blue and boundless edge,
to a sill of nothingness.

One salt-and-vinegar chip
bag, now empty, blows into
the she-oaks on a small bluff.
Three seagulls bob on the tide.
There used to be crayfish here.

Nothing has been reported
to the police yet, at all.
And the years blow past as wind.

THE SHINING GIFT

At this turning-over season
when the pink heath flares
another quick reading of it all
goes through me like the whip
of an arabesquing swallow.

Mind you, though,
every once in a long while
tin evil bumps back,
disconcerting element
like a tooth in a glass of water
which catches your sleepy eye
when you rise to wake the child
and first remember
the glittering coin
you or a coy tooth fairy
meant to leave in that lit glass.

But after all your indigo sleep
(ablaze with free theatre)
there is no money at all
to pay ravenous evil off,
quick as your first cough.

But yes, oh yes, there are
intimations of petalled grace,
the good moment flying
into your loaf of bread;
just when a light-shaft
leans in on the study carpet
you have felt warmth for a mo
come in on the short wave.

Cloud bravadoes along,
an iris telegraphs
and the frantic honey-eaters
tweet and scuffle outside,
all cogged in reality.
Through, over or past them
you rise above your falling
discerning at last
that all of it after all
is your paintblue blessing,
akimbo, sprawled,
exactly the size of your life.

One lifespan is a gift
slowly being unwrapped
in the milky April morning.
The coloured paper crinkles.

A TRANSIENT

And then I see us falling
in a great arc, out through blind space,
never to touch again
the points we passed through
back in those blue days
when we fell in gravel schoolyards
or picked a scab off a knee,
back when
my mother came in on tiptoe
to give me a goodnight kiss
between sandman and nightmare:
a gentle valediction
that was not the kiss of death,
nor yet the final touch
as we are thrown for good
out on our boundless, elliptical
journey through cloudy self
onto lightyears of unmitigated terror,
where language is not
and the veins flow out to sea.

OXFORD POETS

Fleur Adcock

James Berry

Edward Kamau Brathwaite

Joseph Brodsky

Basil Bunting

W. H. Davies

Michael Donaghy

Keith Douglas

D. J. Enright

Roy Fisher

David Gascoyne

Ivor Gurney

David Harsent

Anthony Hecht

Zbigniew Herbert

Thomas Kinsella

Brad Leithauser

Derek Mahon

Medbh McGuckian

James Merrill

Peter Porter

Craig Raine

Christopher Reid

Stephen Romer

Carole Satyamurti

Peter Scupham

Penelope Shuttle

Louis Simpson

Anne Stevenson

George Szirtes

Grete Tartler

Edward Thomas

Anthony Thwaite

Charles Tomlinson

Chris Wallace-Crabbe

Hugo Williams